# Disappearing
# Giants

## The North Atlantic Right Whale

### SCOTT KRAUS and
### KENNETH MALLORY

BUNKER HILL PUBLISHING

IN ASSOCIATION WITH

NEW ENGLAND AQUARIUM

First published in 2003 by Bunker Hill Publishing Inc.
285 River Road, Piermont, New Hampshire, 03779, USA

10 9 8 7 6 5 4 3 2

Library of Congress Cataloguing in Publication Data
is available from the publisher's office

ISBN 978-1-59373-004-8

Designed by Louise Millar

Printed in China by Jade Productions

*The New England Aquarium's thirty-foot research boat,
the* Nereid, *is smaller than most of the whales it encounters.*

The **New England Aquarium** opened in 1969 to present, promote, and protect
the world of water. In addition to its programs and exhibits, enjoyed by more than
1.3 million visitors every year, the Aquarium's conservation and research projects
are among its most important initiatives. The Aquarium's Right Whale Research
Program is the longest continuing program devoted to saving the right whale.

*for Jackie Ciano*

On Sunday, January 26, 2003, the New England Aquarium, the Wildlife Trust, and the whale conservation community lost a dear friend and colleague, Jackie Ciano. She and three of her colleagues were lost at sea in the course of an aerial survey of North Atlantic right whales off Florida. Jackie was born in Arlington, Massachusetts, started as a marine mammal trainer at the New England Aquarium in the 1970s, and moved to whale research as her passion for marine conservation grew. She studied whales around the world, in the U.S., Canada, Denmark, and Norway. With her vibrant personality and boundless personal commitment, Jackie is remembered as a good teacher and problem-solver. Jackie Ciano was dedicated to saving right whales from extinction, and we take some consolation in knowing that she was doing what she wanted to do. We dedicate this book to her memory.

*One of the stories of St. Brendan, who lived in sixth century Ireland, includes his erecting an altar on the back of a sleeping whale, shown here. Whales have appeared as monsters and saviors throughout their early history.* The Whale Assists in the Discovery of New Worlds, *Reproduction "from Philopono's Nova Typis 1621."*

# Introduction

Whales and dolphins have captured human imagination since the beginnings of recorded history. Their images have adorned Mediterranean coins, embellished panoramic frescoes, and animated Norwegian stone etchings dating back to at least 2200 B.C. Greek astronomers included the constellations Cetus and Delphinus (the whale and the dolphin) in documents as old as 366 B.C., and legends about the interactions of humans and dolphins date earlier still.

Among the first evidence of hunting whales is an engraving on a sandstone wall in Korea, possibly from as early as 6000 B.C. The Alaskan Inuit were probably whaling from 1500 B.C. Commercial whaling for right whales began in the late 900s in the Bay of Biscay, and by the 1500s, there was a well-established whale fishery throughout most of the North Atlantic.

For thousands of years, no one knew whether whales and dolphins were fish or mammals. Early British and Spanish decrees refer to the King's rights with regard to the "royal fishes" that included sturgeons, dolphins, and whales. Aristotle was the first scholar to describe many of the features of whales and dolphins, including their warm-blooded nature. Countless writers turned

*This fresco from the Minoan Palace of Knossos on the island of Crete displays reverence for dolphins from earliest times.*

whales into sea monsters, with exaggerated features (the Romans Pliny and Galen, and the German Gesner, who appears to have invented the unicorn!). But the emergence of true whale "scholars," i.e., those more excited by anatomy and biology than by profits, appeared in the 1600s. In 1693, a British scholar named John Ray reported on the true mammalian nature of what were now referred to as cetaceans (from *Cetacea*, their order name in scientific nomenclature).

Today, whaling is a relatively minor factor among the threats that face whales and dolphins around the world. Coastal development, overfishing, pollution, and mineral extraction are all having a powerful effect on their habitats and their sources of food. Despite protection by United States legislation—the Marine Mammal Protection Act (1972) and the Endangered Species Act (1973)—habitat protection for cetaceans is not well established. Research and conservation organizations such as the New England Aquarium, the Center for Coastal Studies, and the College of the Atlantic are all working to change attitudes and improve understanding about the ocean and to find ways to minimize human impacts upon it. In the long term, only by protecting the ocean can we preserve the animals and systems upon which they (and we!) all depend.

*A colored aquatint called* The Whale Fishery *shows the crew of an early nineteenth-century whale boat attempting to harpoon a right whale.*

*The North Atlantic right whale tail, or fluke, is one of the features that help researchers identify the species.*

As fellow mammals, we see mirrors of ourselves in whales—there is a social kinship and shared history. Among the 76 or so species of whales living today is the biggest baby ever to grace our planet, the 30-foot-long blue whale, whose mother's milk is so caloric, its newborn baby weighs 21 tons by the time it is seven months old. Among their legion is the sperm whale, so perfectly designed for diving that it can safely pursue squid several miles deep into the abyss. Other cetaceans such as bottlenosed dolphins form remarkable societies similar to primates in their complexity. And, finally, we include the subject of our book, the massive, graceful, still mysterious North Atlantic right whale. Never has a whale come so close to extinction and still had a chance of survival. Our wish to you, gentle reader: "May your grandchildren live to see one."

7

# What's in a whale

Whales are marine mammals, like seals, walrus, and sea cows called manatees. Though they have lost most of the outer coat of hair that is characteristic of most mammals, whales conform in all other respects. They are warm-blooded vertebrates that give birth to live young, which they suckle with nutritious milk.

All whales come from two major groups, toothed or baleen whales. Toothed whales such as orcas (killer whales), bottlenosed dolphins, or harbor porpoise, use their teeth to consume favorite prey, including fish, squid, or in the case of orcas, even seals or large whales. Baleen whales such as right and humpback whales, filter fishes and tiny floating animals called plankton with long racks of bristly plates called baleen. Up to 300 plates of baleen hang side-by-side from the inside of their upper jaws, like a comb on the inside of the mouth. Baleen, also called whalebone, is made of keratin, the same substance as our fingernails, and each plate is fringed in the inside of the mouth to create a large sieve. The fact that a blue whale is a baleen whale is yet another great irony of the whales. The largest creature on Earth feeds on animals that are often not much larger than a matchstick.

*Orca teeth*

*Minke whale baleen*

*Whitesided dolphin (toothed whale)*

*Breaching humpback (baleen whale)*

*Orca or killer whales (toothed whale)*

# Whale Evolution

The story of how cetaceans first evolved is unique among all mammals (cetacean is another name for whales and dolphins, based on the Latin root name "Cetus," meaning whale). Although their fossil record is still very incomplete, cetaceans are thought to have first developed about 50 million years ago. They appear to have evolved from small-hoofed, wolflike land mammals known as mesonychids (mes-o-NICK-id), close relatives of today's hippos, camels, and horses. Scientists believe they lived in the coastal zones of the ancient Tethys Sea where they foraged on the rich sources of food they found in shallow water near land. They eventually evolved into animals better adapted to marine life. The vestiges of this transition from land to sea can be found in all cetaceans—they all possess remnants of the pelvic girdle, which supported hind legs. In some unusual cases, portions of hip and leg bones have been found where legs used to be!

The first confirmed cetaceans were fully aquatic creatures called archaeocetes (ARK-ee-oo-seats), ranging in size from about 9 to 60 feet (3 to 20 meters), depending upon the species. Most of these primitive whales still had hind limbs in the form of flippers, although scattered fossil evidence indicates that these hind "legs" had disappeared by 40 million years ago. During this same period, the nostrils of this class of animals slowly evolved toward the top of the head from the tip of the snout. For an ocean dweller, the selective advantages of breathing from the top of your head are obvious, and the fossil record documents this remarkable example of evolution over time quite well.

The split between "toothed" whales (Odontocetes) and baleen whales (Mysticetes) occurred between 25 million and 35 million years ago. While the limited fossil record suggests the older date, information based upon a constant rate of mutation in DNA favors a more recent divergence. The right whales (Balaenidae) are collectively believed to be among the oldest representatives of all baleen whales. Details on the evolution of this entire group must wait for the creation of a more complete fossil record, and on advances in molecular (genetic) techniques. Regardless, multiple species of cetaceans have been recorded through the history of their existence, and today there are about 76 species of whales and dolphins.

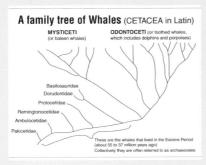

### A family tree of Whales (CETACEA in Latin)

**MYSTICETI** (or baleen whales)

**ODONTOCETI** (or toothed whales, which includes dolphins and porpoises)

Basilosauridae
Dorudontidae
Protocetidae
Remingtonocetidae
Ambulocetidae
Pakicetidae

These are the whales that lived in the Eocene Period (about 55 to 37 million years ago) Collectively they are often referred to as archaeocetes

*A reconstruction of Pakicetus based upon their skeletons. Illustration by Carl Buell, and taken from http://www.neoucom. edu/ Depts/Anat/Pakicetid.html. Skeletons of Pakicetus (a) and Ichtyolestes (b) are ancestral forms of the modern cetaceans. Scientists have long known that cetaceans (whales, dolphins, and porpoises) descended from four-footed land mammals. Cetaceans still have some features of land mammals; they use lungs to breathe air and give birth to young that are nursed by milk produced by the mother. Modern cetaceans cannot live on land, and look very different from land mammals in most respects.*

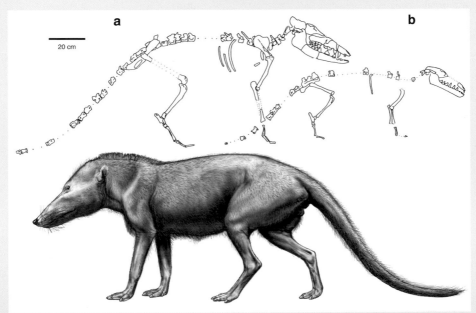

a

b

20 cm

# The Intelligence Question

Whales are superbly adapted for ocean life, but are they really as smart as people would have us think? Researcher Margaret Klinowska offers the following observations. There is no question that the size of a whale's brain can be immense.

Compared to the weight of the average human brain—3.7 pounds—the sperm whale brain tips the scale at an astounding 20.2 pounds. Right whales, by comparison, have brains that only reach about 6.8 pounds. But does a brain's size and weight correlate with intelligence? Or do large bodies require large brains to control them? Does size of the brain relative to the size of the rest of the body really give a clear indication of intelligence? The more we learn about intelligence, the more complex it becomes, and the more scientists are realizing that no single approach to defining intelligence will do.

There is little doubt that toothed whales behave in ways that we typically call intelligent. Like dogs, dolphins can respond to a host of commands, can be trained to respond to behavioral cues, and can even be taught rudimentary language syntax. Baleen whales, on the other hand, behave in ways more equivalent to herdlike animals, i.e., cows or deer—but measuring intelligence in one of these creatures is a daunting task. Until we have a better understanding of what intelligence is, maybe this is the wrong question. These animals have adapted to their lives in the ocean in ways that are remarkably sophisticated. Still, there is no evidence that dolphins could compose symphonies or communicate with us in a language of a higher order than the one humans use.

*In studies of animal intelligence, dolphins excel, as shown by their ability to follow commands and learn new skills.*

# The North Atlantic Right Whale:
## *The Fight against Extinction*

North Atlantic right whales (*Eubalaena glacialis*) are the most endangered large whales in the oceans today. Fewer than 350 are left in breeding and feeding grounds that extend from Nova Scotia to the Gulf of Mexico. Survivors of hundreds of years of commercial exploitation, the right whales we see in the ocean today are barometers for the plight of whales in the twenty-first century. Their story is also a cautionary tale about humans and ocean wildlife, providing insight into other species of whales and dolphins alive today.

The right whale got its name because it was the "right" whale for people who hunted whales. It swam slowly at the surface, so it was easy to catch and kill. Once killed, it floated—making retrieval easy—instead of sinking out of sight. When it was boiled, the right whale's foot-thick layer of blubber produced as many as 70 barrels of lamp and lubricating oil. Its baleen was used for commodities such as corsets and buggy whips.

For over 900 years, beginning about the year A.D. 1000, whaling nations from Europe and America hunted North Atlantic right whales until they had almost completely disappeared. By 1935, when they were given international protection as an endangered species, some scientists suspected that there were fewer than 100 right whales left in the North Atlantic Ocean. Most thought the right whale was doomed to extinction.

*New Bedford Harbor, waterfront scene. Whalers Meteor and Sunbeam; casks sit on wharf. 8"x10" glass negative, c. 1870s.*

## Table 1. The most endangered populations of baleen whales.

| SPECIES | POPULATION | ABUNDANCE | LACK OF RECOVERY DUE TO |
|---------|-----------|-----------|------------------------|
| **Northern Right Whale** | eastern North Pacific | tens? | recent hunting |
| | western North Pacific | low hundreds? | recent hunting |
| | eastern North Atlantic | extinct? | wiped out by hunting ca. 1900 |
| | western North Atlantic | about 300 | anthropogenic mortality? |
| **Bowhead Whale** | Davis Strait/ Hudson Bay | 450+ | recovery status unclear |
| | Spitsbergen/ Barents Sea | tens? | intensive historical whaling |
| | Okhotsk Sea | low hundreds | recent hunting |
| **Gray Whale** | Korea/ Okhotsk Sea | 250? | recent hunting |
| **Blue Whale** | all except eastern North Pacific and perhaps eastern North Atlantic | unknown | recovery status unclear |

# Where have all the Right Whales gone?

## *Early Rediscoveries*

Hope for the recovery of North Atlantic right whales began in the 1960s. When Woods Hole scientists Bill Watkins and Bill Schevill began a search to record sounds made by large whales in Cape Cod Bay, off the coast of Massachusetts, they were surprised to find a number of right whales.

It wasn't until 1979, however, that University of Rhode Island scientist Howard Winn got a clearer picture of right whale population numbers. During large-scale aerial surveys for all marine mammals and sea turtles along the northeastern coast of the United States, Dr. Winn recorded more than 60 right whales.

*Courtship groups like the one shown here from the Bay of Fundy also occur in and around the coast of Massachusetts.*

## Bay of Fundy

In 1980, a small group of researchers, led by Scott Kraus and John Prescott from the New England Aquarium, were asked to conduct a survey of the marine mammals in the Bay of Fundy, an area between New Brunswick and Nova Scotia. An oil company planned to place a refinery in the bay. The government wanted to know what impact the building of a large tanker port might have on the surrounding wildlife. To everyone's great surprise, 26 different right whales were discovered during these surveys, including four mothers with their calves. The port and refinery were never built.

Kraus's discovery of a group of right whales swimming with their calves was another sign of hope for the species. In the years that followed, research teams from the New England Aquarium returned to the Bay of Fundy each summer to study the whales. They learned that the bay is the summer and fall nursery ground for most right whale mothers and their new calves in the North Atlantic.

In the bay, strong tidal currents mix cold water and nutrients into a rich soup that supports large amounts of right whale food, including copepods, which are their favorite food. In the Bay of Fundy, swarms of these

*The Bay of Fundy with its sheer cliffs, rocky shore, and fog-bound coastline is an area with a strong tidal surge and an upwelling of abundant nutrients that sustain the food right whales depend on.*

animals form dense underwater clouds. The clouds are so thick that the right whales just open their gaping jaws to feast.

The New England Aquarium research team's discovery of right whales in the Bay of Fundy had revealed the northern, or "summer," end of the right whale's migration, but Kraus still didn't know where the southern end might be. It took a look back over a hundred years to give them a clue.

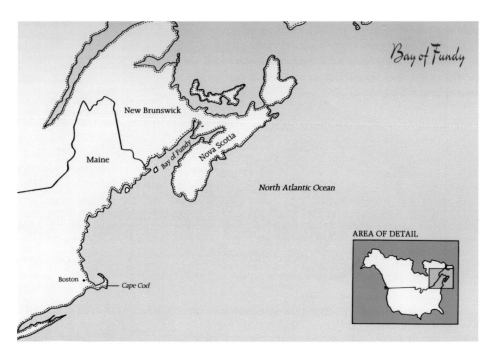

*Two of the critical habitats North Atlantic right whales depend on are the Bay of Fundy between New Brunswick and Nova Scotia, and Massachusetts Bay off Cape Cod. See page 43 for all critical habitats.*

# Feeding

North Atlantic right whales migrate into high latitude waters to feed, usually from Massachusetts north to Nova Scotia in the Western North Atlantic Ocean. Their primary prey is the copepod *Calanus finmarchicus*, a microscopic relative of shrimp and crabs whose adult size is about the size of a grain of rice, but other small zooplankton (small floating animals) such as barnacle larvae and different copepod species, are eaten at times. The whales filter their food by swimming continuously with their mouths open, either at the surface (skim feeding) or at depth. Feeding sessions at the surface can last for hours. When feeding at depth (down to over 600 feet, or 200 meters), dives of up to 20 minutes or more may be repeated for hours. As many as 270 finely fringed baleen plates, up to 8 feet long, hang from each side of the upper jaw, allowing the whales to filter small zooplankton from the water. The whale only opens its mouth and exposes its baleen when the concentration of plankton animals in the water reaches a critical mass. Dr. Stormy Mayo of the Center for Coastal Studies reported that right whales in Cape Cod Bay did not skim feed unless there were more than 1,000 zooplankton organisms per cubic meter of water (about the amount of water in a bathtub or a dishwasher). Zooplankton usually occurs in highly concentrated "patches" in the water column driven by tidal and other currents. High densities of copepods are found in all spring, summer, and fall right whale habitats, but have not been found in the wintering grounds off the southeastern United States. Scientists believe that right whales may consume up to 2,000 pounds of zooplankton per day.

*A right whale at the water's surface gulps in water and plankton, trapping it on its baleen plates, shown here in this image (opposite). For such a large animal, the size of its food, called copepods and shown here swarming in the jar, is surprisingly small (above).*

## *History Teaches a Lesson*

In 1983, researchers Randy Reeves and Edward Mitchell found a missing piece of the puzzle. Searching through the logbooks of nineteenth-century whaling ships, they discovered that a whaling schooner from New Bedford, Massachusetts, named Golden City, had pulled into Brunswick, Georgia in the winter of 1876. It went there to remove its cargo, which included whale oil and whalebone from humpback whales taken near the Bahamas. But that year, instead of continuing farther north, as it usually did, the Golden City remained around Georgia for a couple of months. While it was there, the crew captured a single right whale.

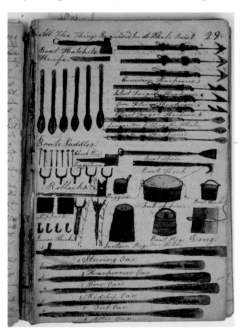

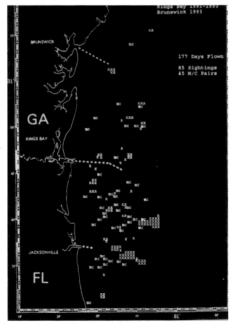

*Logbook aboard Dr. Franklin (Sailing Vessel) of Westport, MA, 1856-1859, North and South Atlantic grounds.*

*The warmer waters of Georgia and Florida attract right whale mothers as a place to give birth.*

They must have seen more, because the following year a second whaling vessel appeared in Brunswick, and by 1880, five whaling vessels had made Brunswick their winter headquarters. Records showed that at least 25 to 30 right whales were killed there between 1876 and 1882.

In modern times, a few right whales have washed up dead along the coast of the southeastern United States, but there were few sightings of live whales in the area. Even so, if right whales had appeared off the coast of Georgia during the winters over a hundred years ago, as the whaling logs suggested, there is a chance they might still do so today. Perhaps the right whales Kraus and his team of researchers discovered that summer in 1980 had traveled there during their seasonal migrations. But to prove it, researchers would need to survey the area, and they would need a way to recognize and follow individual right whales.

*Until researchers did exhaustive aerial searches in Southeast U.S. coastal waters, often the only right whales recorded in modern times were those that had washed up on the beach. The baleen plates are visible in the mouth of this dead beached right whale.*

21

## *Hunting Right Whales*

Right whale hunting started as a shore-based fishery by the Basques over a thousand years ago, off northern Spain and western France. By the early 1500s the Basques had expanded their operations to the coasts of Newfoundland and Labrador, and for nearly a century, they hunted right whales and bowhead whales (*Balaena mysticetus*) near the Strait of Belle Isle every summer and autumn. The Basque whalers killed between 25,000-40,000 whales between the years 1530-1610. In New England, right whale hunting started from colonial shore stations during the late 1600s, peaked in the early 1700s, and persisted at low levels until the early 1900s. In addition, the American pelagic (open ocean) whalers killed right whales at several locations around the North Atlantic until nearly 1900.

The discovery of petroleum in 1859 reduced the demand for whale oil as fuel for lamps and as a lubricant. Baleen remained in great demand, however, well into the 1900s, and right whales continued to be valuable prey. Whaling for right whales didn't stop until 1935, when the League of Nations finally gave it international protection. We will never know the original population size of North Atlantic right whales before hunting, although it was probably several thousand animals. It is likely, however, that by 1900, only a few dozen remained in the western North Atlantic.

*This colored line engraving, entitled* A Whale Female and the Windlais Whereby the Whales Are Brought on Shore, *shows a right whale being processed on shore in the year 1619.*

22

*Hoisting on the lower jaw of a sperm whale; cutting in aboard the California (Steam Whaler), 1903. Photograph by Marian Smith.*

## The Who's Who of Whales

With a stocky, rotund body that lacks a dorsal (back) fin, right whales display individually distinctive patterns of thickened skin, called "callosities," on their heads. Using the callosities as well as other field markings, researchers can tell each whale apart, just the way humans can tell one another apart. One of the most important tasks of the Aquarium's yearly surveys in the Bay of Fundy has been to collect right whale "mug shots." Photographs and drawings showing callosities, scars, and other distinctive markings have been combined to produce identification files for each individual whale. Together with other research teams studying whales, the Aquarium's researchers began assembling a right whale catalog in the 1980s, sharing identification data that provides information on individual whales and their movements.

All North Atlantic right whales are numbered and cataloged at the Aquarium. Some whales also are named for distinctive features, so that researchers can easily remember unique whales while at sea. For example, "Stars" and her mom "Stripe" were two of the earliest entries in the catalog, one named for small starlike marks on her head, and the other for a large stripelike scar across her nose. "Snowball" has a circular scar on the side his head, as if he had been hit with a snowball. Stumpy, Droopy, Admiral, Smoothie, Starry Night, Kleenex, Necklace, Baldy, and Spitball were just a few of the other names that whales acquired during the early years, usually from easily identified marks.

*Right whale mug shots show markings called callosities that are unique to each individual whale. Researchers use these raised calluses to identify individual animals in their yearly migrations.*

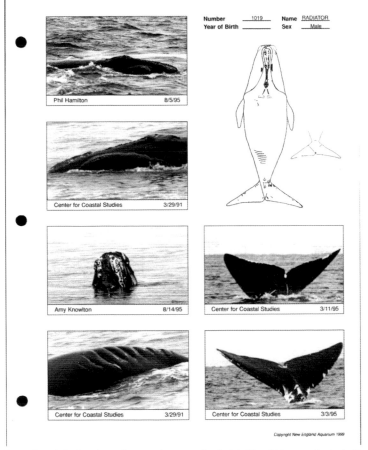

Number ___1019___ Name _RADIATOR_
Year of Birth _____ Sex ___Male___

Phil Hamilton 8/5/95

Center for Coastal Studies 3/29/91

Amy Knowlton 8/14/95

Center for Coastal Studies 3/11/95

Center for Coastal Studies 3/29/91

Center for Coastal Studies 3/3/95

Copyright New England Aquarium 1999

*The Right Whale Catalog calls attention to each whale's identifying marks, including callosity patterns, scars, scratches, and belly patches.*

25

# Research Methods

Before the 1700s, scientists studying whales relied upon animals that had washed ashore. The bodies were in various states of decomposition, and the subsequent reports led to descriptions of monsters of various sizes and shapes bearing little resemblance to the real creature. In the 1800s and 1900s, students of whales put to sea aboard whaling ships. While this yielded a lot of information about anatomy, it told little about the animal in the wild. The science of whales has come a long way in the last 30 years, now yielding a wealth of data on almost every aspect of an animal's life, all without the need to kill it.

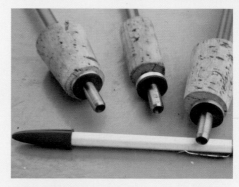

*DNA darts have a hollow tube at one end that collects samples of skin and blubber. The cork assures the dart will float for scientists to collect for study. In the image opposite, a DNA dart is visible on the back of a right whale.*

26

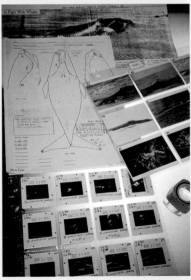

Researchers use a crossbow to send the DNA dart into the blubber of a right whale. The dart design assures it will only penetrate superficially in the whale's flesh. Records are kept on each whale's life history, including images of each whale as seen over the years.

27

# The Right Whale Calving Grounds

As luck would have it, in the fall of 1983, Dave Mattingly, a professional airline pilot, called Kraus's office to volunteer his services. He was able to organize a group of pilots with private aircraft and flying time to carry out the survey off the coasts of Georgia and Florida. Most people thought they were crazy. No one had searched for right whales in the area since whalers left the region nearly 100 years earlier. When Dave told his friends he was looking for right whales off the coast of Georgia, they looked at him as if he had said he was going to downtown Atlanta to look for dinosaurs.

*Spotter planes such as this one help alert commercial vessels that right whales are in their shipping lane.*

The survey took place in February 1984. Each plane had four people on board. One person flew the plane, one took notes, and the other two acted as observers, whose jobs were to find whales. The planes flew in a pattern that crossed back and forth in lines, or "tracks," over a section of the ocean, as if mowing a lawn. If there were any whales to be seen, this method would increase the chances of finding them.

It wasn't until the second day that surveyors saw anything except turtles and sharks. Halfway through that day's flights, pilot Jon Hanson reported something in the distance. When the team flew in for a closer look, it turned out to be a mother and a baby right whale. By the end of three days of flying, they had counted 13 right whales, including three mothers with their newborn calves. They were all swimming in coastal waters near the border of Georgia and Florida. One of the mother whales was old friend Stripe. With her was her fourth calf "Forever." From her size, "Forever" might have been about two months old.

After that first survey, the Aquarium research team and the volunteer pilots repeated their efforts for five years. Based on this work, researchers learned that the

*Mother and calf loll at the surface along the Florida coast in preparation for the long swim north for food in the spring and summer.*

coastal zone of Georgia and Florida is the primary calving ground for all right whales in the North Atlantic.

From December to March of each year, female right whales give birth off the southeastern coast of the United States. Mothers and calves stay close together, usually apart from other mothers and calves. After its birth—tail first and underwater—the newborn calf does little except eat, play, and rest. It feeds from two nipples hidden in slits along its mother's belly. The milk provides all the nourishment the baby needs. Whale milk is so rich in fat that a newborn right whale may gain a ton—2,000 pounds—in a little over a month.

Today, teams of researchers from the Aquarium and the states of Georgia and Florida fly surveys over the calving ground every good weather day, in order to provide information on the whales' locations to all mariners in the region. The U.S. Navy coordinates sightings in Jacksonville, Florida, and transmits them to all shipping and military traffic entering or departing through the calving ground. These "Early Warning Surveys," although only effective during daylight hours, have alerted ships to the presence of hundreds of whales in their paths over the last ten years. Ships on alert have frequently changed course to minimize their chances of hitting right whales, and this system can be credited with saving the lives of several dozen right whales.

**29**

## *The Long Swim*

With the help of the right whale catalog, Aquarium researchers have made many photographic matches between whales in the calving grounds and whales seen in the Bay of Fundy and in other places along the North American coast. These matches and surveys, combined with the work of other researchers in Massachusetts and Rhode Island, have begun to provide a picture of the yearly migrations of the right whales along the Atlantic coast.

Many different species of whale migrate, or travel, between winter breeding and calving grounds and summer feeding grounds. Migration is one way in which animals have adapted to changes in the seasons, and it is common in birds, herding mammals, whales, and many fishes. In whales, most summer migrations move into productive northern waters, where the food is abundant. In the winter, pregnant females move southward, so that when they give birth, their calves can spend their first weeks of life in warmer water.

With the arrival of warmer weather and increasing daylight during the spring, mothers and calves begin a 1,400-mile journey north along the Atlantic coast. They stay close to each other because the calf is still very dependent on its mother's milk. But if they do get separated, we think they use underwater sounds to help them keep in touch. Although the exact path that mothers and their newborn calves take along the coast is unknown, sighting records and satellite tracking data suggest that this migration occurs within 30 miles of the eastern seaboard.

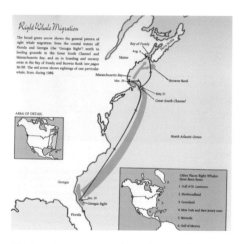

*The broad green arrow shows the general pattern of right whale migration: from the coastal waters off Florida and Georgia (the "Georgia Bight") north to feeding grounds in the Great South Channel and Massachusetts Bay, and on to breeding and nursery areas in the Bay of Fundy and Browns Bank. The red arrow shows sightings of one particular whale, Stars, during 1989.*

*A right whale named Spitball is seen here off the coast of the southeastern United States (bottom right). Researchers make detailed reports of each right whale sighting (bottom left). A right whale "blow" (top right) or breach (top left) helps scientists locate and follow specific animals.*

# Right Whale Senses

Right whales appear to produce sounds only for communication. Northern right whales seem to have a complex repertoire of up, down, constant, and high pulses—all associated with social activity and mating. There is no evidence of echolocation (biological sonar) in any baleen whale that is equivalent to the proven abilities of toothed whales. Right whales are known to use well-developed hearing capabilities to find mates, and probably use passive listening to avoid predators (like killer whales) and vessels, and possibly for long-range navigation.

Vision appears to be the primary sense right whales use to find prey and for close-range navigation. Although anatomical studies have only been done so far on eyes from other baleen whales, all whale vision appears to be similar. Whale eyesight is well adapted to the lower light levels found below water, with fairly high resolution. Based on anatomical analyses of olfactory lobes in the brain, a whale's sense of smell and taste is limited. While there is some evidence that whales have the cellular machinery that would allow them to use magnetism for navigation, there is no proof that right whales do so. Right whales are highly sensitive to touch, and engage in a large amount of close contact with one another, both in mothers with calves, and in courtship.

*A right whale eye is about the size of a grapefruit. Right whale vision is adapted to the low levels of light found below water. The right whale blow holes produce a V-shaped spout that distinguishes it from whales that produce a single spout.*

## *Springtime in Massachusetts*

By April, and sometimes earlier, the first of the right whale mothers and calves arrive in waters off the Massachusetts coast. Throughout the winter, and during their journey north, the female whales have been living off the fat they store in their blubber. But as they reach cooler waters, they begin to feed. Researchers have observed right whales feeding in the Great South Channel—a deep-water passage between the tip of Cape Cod and a productive underwater fishing bank 60 miles to the east called Georges Bank—and in Massachusetts Bay.

In Cape Cod Bay, Center for Coastal Studies scientist Dr. Stormy Mayo takes advantage of the seasonally resident right whales to study the intricacies of feeding behavior and energetics. His studies, and those of University of Rhode Island researcher Dr. Robert Kenney, have provided extensive information on the remarkable links between oceanography, productivity, copepod behavior, plankton aggregations, and right whale feeding.

*A favorite pastime in Massachusetts Bay is skim feeding shown by the right whale here using its open gape to swim through swarming plankton, trapping them on its baleen plates. In their feeding habits, courtship groups and breaching animals can be a common sight.*
*(Left) Scott Kraus (right in photo) greets a North Atlantic right whale in Massachusetts Bay.*

**33**

## *Summer in Canada*

By the end of July many right whales reach the Bay of Fundy, where they continue their feast. The whales observed in the Bay of Fundy remain there sometimes until late October. In 1986, six years after the discovery of right whales in the Bay of Fundy, researchers had solved part of the puzzle of the right whale's migration. They knew that many whales spent their summers feeding in the Bay of Fundy and the springtime in the waters of Massachusetts. They also knew that most right whale mothers about to give birth spent their winters in the calving grounds off the coasts of Georgia and Florida. But part of the puzzle was still missing: Where did the females without calves and most of the male whales go?

In August 1986, researchers sailed from Yarmouth, Nova Scotia, aboard the fishing boat Melissa Marie, under the command of fishing master Delphis Doucette. The

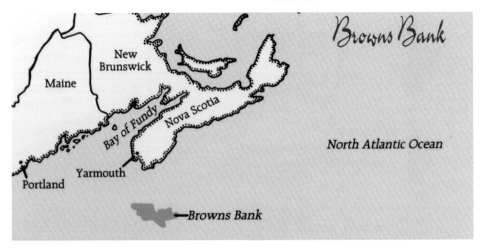

*The waters off northern Maine towns like Lubec (shown left) and off Nova Scotia at Browns Bank (shown above) are critical habitat areas for the North Atlantic right whale.*

destination was an area near Browns Bank, a shallow fishing ground south of the southern tip of Nova Scotia. Howard Winn of the University of Rhode Island had reported right whale sightings from this area in the early 1980s, and the Aquarium team wanted to find out whether it might be a summer feeding ground for adult right whales.

Captain Doucette wasn't convinced—he had fished in the area before, and had never seen many whales there. The next two days left him shaking his head in astonishment. By the end of the second day of their survey, when it began to get too dark to take pictures, Kraus and the team had photographed nearly 70 different whales. They tried to make an accurate count, but they were overwhelmed. In this relatively small area, about four miles in diameter, they estimated that there were over 100 right whales.

The whales were diving and feeding on the rich soup of copepods in the cold water. More exciting, however, were observations of 19 "courtship groups," including one with at least 20 whales in it.

**35**

# Mating and Reproduction

Right whales mate with many different partners during their reproductive lives. The reproductive cycle is over three years long, so less than one-third of the adult females may be receptive to males each year, making them highly sought after. Mating takes place in the midst of large courtship groups, where females appear to call males to them, and then make mating difficult, either by swimming away or by lying on their backs. Males compete for access to the female through active pushing and by displacing one another, and multiple copulations have been observed. Males also probably employ sperm competition—male testes are both the largest in the world >1764lbs (800kg) and the largest relative to body size in the baleen whales. Courtship groups can last up to six hours, and may comprise as many as 40 animals, usually with only one or two females.

Calves are born in the winter months at a length of 13-15 feet (4-5 meters). Cows give birth to a single calf every 3-5 years. The length of gestation is estimated to be about 12-13 months. Infant right whales grow to between 24-28 feet (8-9 meters) by the end of their first year, eventually reaching 45-55 feet (15-18 meters) and 40-50 tons as full-grown adults. The lactation period is 10-12 months, although calves occasionally stay with their mothers up to 17 months. The mean age at first calving is about ten years.

Recent assessments of North Atlantic right whale reproduction show that in the late 1900s mature female right whales have been producing fewer calves per year. Mean calving intervals have increased significantly from 3.67 years (1980-1992) to over five years during the period between 1996 and 2000. The population growth rate is substantially lower than southern right whale populations off Argentina and South Africa. The causes of poor reproduction are unknown, but may be due to low genetic variability, or oceanographic or climatic changes that have reduced food. Other problems may include diseases, biotoxins from red tides, and the sublethal effects of pollutants.

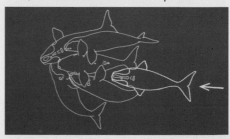

*Right whale males compete for the attention of a single female, jostling and crowding one another to try to mate with her.*

# Social Structure and Behavior

Right whale social structure is poorly understood. Individual whales may be seen alone at some times of day, and also with one or more groups later that day or on other days. When groups of whales are found within a few kilometers of one another, it is most likely in response to concentrations of food and is not social behavior comparable to pods of dolphins. The most tightly linked social pairing is mother and calf. They can remain within one body length of each other for the first six months after birth. Weaning occurs at 10-12 months, and mothers and their offspring are rarely seen together again.

Breaching behavior (leaping from the surface) and lobtailing (slapping the water with the flukes) occur frequently in this species. Such behaviors may help right whales locate each other, especially when increased sea surface noise limits the range at which vocalizations might otherwise be heard.

*Top to Bottom: Breaching or jumping out of the water may be a way for right whales to attract attention or spread alarm. Mothers and calves bond together for the first years of the newborn's life. Lobtailing, or raising the tail in the air and slapping it on the water, may be another way to tell other whales where they are.*

37

# The Lost Right Whales of the North Atlantic

Researchers still don't know where the males and females without calves go year after year during the winter. Most of the North Atlantic is a wilderness of ocean, untraveled except for the fishing and cruising boats that pass from time to time. Imagine a population of right whales numbering only a few hundred scattered over thousands of square miles of ocean and you have an idea of how difficult a task it will be for researchers to find them.

If the whereabouts of right whales in the winter remains unsolved, so too are there questions about a possible unidentified summering ground, hinted at through genetics and surveys by Dr. Brad White of Trent University and Dr. Cathy Schaeff of American University. Their work suggests that a full third of the population go to a summer location other than the Bay of Fundy or Nova Scotian Shelf. Historical sightings compiled by Reeves and Mitchell point to potential offshore summer right whale habitats, but surveys will be needed to determine if animals are currently using them.

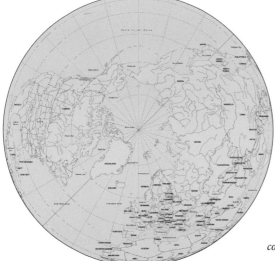

*Every year some of the right whale females and males disappear to some as yet unknown location. Researchers think the waters between Greenland and Iceland may be one such undiscovered habitat (see map to the left). They have already begun to plan a series of expeditions there to look for callosity-covered whales like the one pictured opposite.*

# Threats to Survival

Despite protection from whaling for more than 65 years, the population of North Atlantic right whales is growing very slowly, with perhaps no more than 350 alive today. Collisions with ships and entanglements in fishing gear are responsible for nearly 50% of all right whale deaths. And evidence is mounting that reproduction has been impaired for the last ten years because of pollution, limited genetic diversity, disease, and biotoxins from red tides.

Continued research into the biology of right whales is a crucial part of protecting their future. Without this research, scientists cannot hope to protect their habitat—and without some help protecting their habitat, there is a danger that the North Atlantic right whales may disappear forever from the oceans. If current rates of mortality don't change, this species will be extinct by the year 2200.

*Propeller scars show the danger of collisions with boats.*

*Other threats to right whales are entanglement with fishing lines (top) and collisions with ships (bottom).*

41

# Hope for the Future

Although collisions with ships remain a critical area of concern, scientists and industry managers are all working hard to minimize their impact. It won't be easy. The deep basins of the Gulf of Maine concentrate the food that sustains right whales and the ships that are the primary threat to the whales' survival. Similarly, all ship traffic from Boston and other Gulf of Maine ports to southern destinations must transit through the Great South Channel in order to avoid Georges Bank. In the southeastern U.S., three ship channels track directly through the heart of the right whale calving ground. Collision with ships claims at least one or two North Atlantic right whales annually and may account for a significant reduction in the growth rate of the remaining population as well.

Despite these ongoing threats, however, there are stories of success. In the Bay of Fundy, shipping lanes went directly through the right whale feeding and courtship area for more than 30 years. Thanks to a unique collaboration formed by Dr. Moira Brown and Amy Knowlton of the New England Aquarium with the shipping industry and government of Canada, the shipping lanes now pass slightly to the east of the main right whale habitat. Recently approved by the International Maritime Organization, the revised shipping lanes will greatly reduce risk to right whales without any significant costs to the shipping industry.

*Two right whales at surface side-by-side, and a breach.*

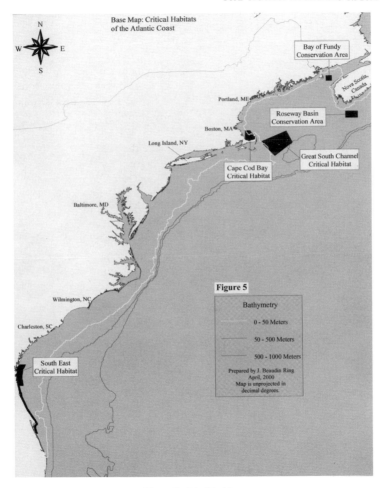

Base Map: Critical Habitats
of the Atlantic Coast

Bay of Fundy
Conservation Area

Nova Scotia,
Canada

Portland, ME

Roseway Basin
Conservation Area

Boston, MA

Long Island, NY

Great South Channel
Critical Habitat

Cape Cod Bay
Critical Habitat

Baltimore, MD

Figure 5

Bathymetry

0 - 50 Meters

50 - 500 Meters

Wilmington, NC

500 - 1000 Meters

Charleston, SC

Prepared by J. Beaudin Ring
April, 2000
Map is unprojected in
decimal degrees.

South East
Critical Habitat

*Map of critical habitats.*

Other urgent work is being done to reduce whale entanglements in fishing gear, which is as threatening to right whale survival as collision with ships. Fixed traps and nets are widely distributed both inshore and offshore all along the coast of North America. But no clear pattern has emerged to guide a management strategy, other than seasonal closures of known whale habitats to risky fishing gear. On the positive side, progress is being made to develop new "whale safe" fishing gear. In the interim, Dr. Stormy Mayo and his team from the Center for Coastal Studies have organized valiant rescue efforts to disentangle whales from fishing gear at sea. Although everyone recognizes that these efforts are not a long-term solution, their work is indispensable in the efforts to reduce accidental kills.

Finally, researchers are increasing their efforts to understand other reasons the right whale population is recovering so slowly. Reduced habitat, biological poisons (biotoxins), and pollution runoff from urban areas are suspect, and so is disease and the danger of inbreeding in a population of animals that is so small. There is no point in saving right whales from shipping collisions and fishing gear if their homes have been lost to the increasing urbanization of the ocean. Whales can certainly tolerate a certain amount of habitat degradation, just as humans can. We don't know, however, how or if these factors are affecting reproduction, feeding, and survival of right whales.

North Atlantic right whales are examples of the failure of humans to share the resources of our planet with other animals through overhunting or destruction of their living spaces. Nevertheless, the outlook for the future of the North Atlantic right whale may be improving. Scientists and managers have identified a number of actions that may bring the right whale back from the brink of extinction. Congressional support for conservation measures is strong, and both the U.S. National Marine Fisheries Service and Canada's Department of Fisheries and Oceans are actively involved in reducing human effects on right whales. All of these efforts will be necessary to save the right whale—but there is a good chance it can be done.

# Pacific Gray Whales—a Success Story

While it's beyond the scope of this story to examine the sad history of man's relationship with whales, the days of uncontrolled commercial whaling appear to be over. Today's threats to whales are less from hunting and more from the inadvertent urbanization of the ocean by humans. To understand this predicament, a look at the story of the North Atlantic right whale (as opposed to the southern right) and the gray whale of the eastern Pacific coast is instructive.

The North Atlantic right whale and the Pacific gray were similarly decimated by the relentless whaling of nearly two centuries. Pacific gray whales make an annual migration thousands of miles from summer feeding waters off the Alaska coast south to birthing lagoons off Baha, California. Yankee whalers discovered this nursery and by the 1900s had wiped out nearly the entire gray whale population. The North Atlantic right whale was the "right" whale to kill because they could often be found slowly swimming at the surface and had huge oil yields. From an estimated pre-commercial whaling population of more than 20,000 whales, the number of gray whales was reduced to a few hundred or so. The North Atlantic right whale fared even worse, going from an estimated 10,000 animals to as few as 50 to 100 individual whales in the 1920s.

*A Pacific gray whale lolls at the surface to breathe.*

*Scammon's lagoon in Baja, California is a gray whale calving ground where gray whales come to give birth and take care of their young. Here a Pacific gray whale peeks at a boat of curious onlookers.*

Today, thanks largely to a hunting ban instituted in 1946 by the International Convention for the Regulation of Whaling, the gray whale has rebounded to numbers even higher than before, to an estimated 21,000 whales. The North Atlantic right whale, on the other hand, even though protected since 1935, is the most endangered "great" or large whale in the ocean, with as few as 300 individuals remaining and no strong indication of an improvement in sight. Why is there such a different result for animals given essentially the same amount of protection?

For Scott Kraus and gray whale expert Steve Schwartz, some of the reasons gray and right whales have responded to protection with such different results are biological. A right whale mother may take four to five years between births compared to the gray whale's two. Another reason for recovery differences is purely statistical. Gray whales began their recovery with several hundred animals, while right whales may have started with as few as 50 whales.

But a third recovery factor, and one which may be key for the future of all whales, is what is being done to protect the places where whales go to feed and breed. The famous Scammon's lagoon is now a gray whale sanctuary protected by the Mexican government. In contrast, the North Atlantic right whales travel in waters that are among the most heavily industrialized coastal zones in the world. Right whale deaths from accidental encounters with fishing gear and shipping, combined with unknown effects of habitat deterioration, have pushed this population into a decline.

# Suggested Reading

Billinghurst, Jane. *The Spirit of the Whale: Legend, History, Conservation.* Stillwater, MN: Voyageur Press, Incorporated, 2000.

Clapham, Phil. *Right Whales.* Stillwater, MN: Voyageur Press, 2004.

Ellis, Richard. *Men and Whales.* New York: Alfred A. Knopf, 1991.

Ellis, Richard. *The Book of Whales.* New York: Alfred A. Knopf, 1985.

Fontaine, Pierre H. *Whales of the North Atlantic: Biology and Ecology.* Sainte-Foy (Québec), Canada: Editions MultiMondes, 2000.

Heyning, John. *Masters of the Ocean Realm: Whales, Dolphins, and Porpoises.* Washington: University of Washington Press, 1995.

Hoyt, Erich. *Seasons of the Whale.* Halifax, Nova Scotia: Nimbus Publishing Limited, 1990.

Katona, Steven. K., Rough, Valerie, Richardson, David T. A *Field Guide to Whales, Porpoises, and Seals: From Cape Cod to Newfoundland.* Washington and London: Smithsonian Institution Press, 4th edition, 1993.

Mann, J., Tyack, P.L., Connor, R., & Whitehead, H. (eds.) *Cetacean Societies: Field Studies of Whales and Dolphins.* Chicago: University of Chicago Press, 2000.

Perrin, W.F., Würsig, B. and Thewissen, J.G.M. (eds.) *Encyclopedia of Marine Mammals.* San Diego: Academic Press, 2002.

Pryor, Karen and Norris, Kenneth, editors. *Dolphin Societies: Discoveries and Puzzles.* Berkeley and Los Angeles: University of California Press, 1991.

Reeves, R.R., Stewart, B.S., Clapham, P.J. & Powell, J.A. *The Audubon Guide to Marine Mammals.* New York: Alfred A. Knopf, 2002.

Russell, Dick. *Eye of the Whale: Epic Passage from Baja to Siberia.* New York: Simon & Schuster, 2002.

Scammon, Charles M. *The Marine Mammals of the Northwestern Coast of North America.* New York: Dover Publications, Inc. 1968.

# Photo/illustration credits